Introduction

To Young Scientists and Their Helpers

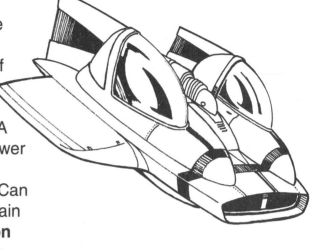

A spaceship goes to the moon. A new type of television comes on the market. Engineers unveil a new jet airplane design. Automobile manufacturers showcase a car with a new type of engine. All of these exciting things are the result of inventions.

When a question is asked, an inquiry has begun. A science inquiry is designed to produce a valid answer to a question asked. Different questions require different designs. In an invention, the question is "Can I create an apparatus (or process) that meets certain specifications?" This book addresses the **invention** design. Other designs include those for collections, observations, models, and experiments.

What is an **invention**, and what purpose does it serve? An invention can be one of two things. First, it can be something or some process that has never been made or done before (for example, the first spaceship, the first car, or the first airplane). The other type of invention is one in which a thing or a process is modified in some way (for example, a better television, a better brake system in a car, or a better mousetrap). Such a changed thing or process is still considered an invention. These kinds of changes occur all the time.

This book helps the young scientist and his or her mentor go through all the steps of creating an invention—from selecting what kind of invention to design and test to displaying the final product at a science fair. A sample invention is developed throughout this book to ensure understanding. (The sample invention involves a child, Alex, who wants to design a paper airplane that flies a certain distance.)

Whether inventing something new or changing something already in existence, the scientist will be combining the skills of **model building** and **experimentation**. Specifications for what the new apparatus or process (invention) can do have to be identified. A model of the invention is built and tested experimentally until the specifica are met.

Model A

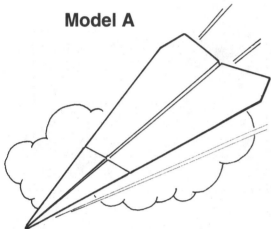

This book has been designed to provide a step-by-ste guide to helping a child invent something. It takes the child through the steps of creating a Log Book, choosing a topic for an invention, gathering information, designing the invention, building and testing the invention, comparing the results of the test with the specifications, and sharing the study. Each step is fully explained and will enable the child to have an interesting and productive time creating his or her very own invention.

There are three roles that the adult helper can play during the study. These involve **mentoring**, **coaching**, and **gophering**. The **mentoring** role involves listening to what the child is asking and helping him or her by asking important questions. You may want to ask questions similar to the following: Do you think that you could get the materials to build that model? Could you test that model during the winter months? What would happen if you try _____? Have you thought about doing _____? Just remember, your role is to guide the child. It is the child who should do the decision-making and the learning.

Coaching involves encouraging the child as well as teaching. You should coach the child in skills needed to carry out the invention study. You should also encourage him or her and provide the support the child needs.

Gophering involves taking the child to see a particular invention, helping him or her find materials needed for the invention, and taking the child to the library. It also involves supporting the child by providing a place to work, build, and test the invention.

Inventions are fun. They enable a child to think, build, and experiment using all of the senses. All of these opportunities help the child grow and develop and understand science. You will be delighted to see a child's self-esteem soar after he or she has invented something that does just what was intended.

Planning Calendar and Log Book

Chapter 1

▦ Planning Calendar

Once the child has decided that he or she wants to invent something, a plan must be made. A plan will help the child know what to do and when to do it.

To create a plan, the child first needs to know when the invention must be completed. Have the child find that date on a calendar and circle it in red. Then, to help the child find out how many days he or she has to complete the invention, the child can number the dates on the calendar backward. The date circled in red is Day 0. The date before that is Day 1. The child can continue numbering to today's date. This lets the child see how many days he or she has to complete the study.

If the child is doing the invention study for a science fair, most science fairs are held in February through April. This is important in the child's planning. If the child wants to build a solar cooker, he or she needs to do it in warm weather on clear days when the sun is visible. Also, remember that some things cannot be studied in certain parts of the country. It would not be feasible to test a new type of surfboard if the child lives in Colorado. Likewise, it would not be possible to test a new kind of snow ski if the child does not live in a snowy region.

It is also important for the child to know that different kinds of inventions may take longer than others to be developed and tested. To help the child decide how long it will take to finish each part of the invention study, have him or her fill out the Planning Calendar on Log It! #1, page 9. (The Log It! pages, when completed by the child, guide the child through a successful invention. They are found throughout the book.) This will be the first page in the child's **Log Book**. (The Log Book is explained in complete detail on pages 4–8.) Remind the child that the filled-in dates are merely suggested times needed for completion.

The Planning Calendar provides the child with an easy way for him or her to see if the invention is on schedule and what tasks still have to be done. When each task is completed, the child should write in that date so that he or she can see what tasks have been accomplished.

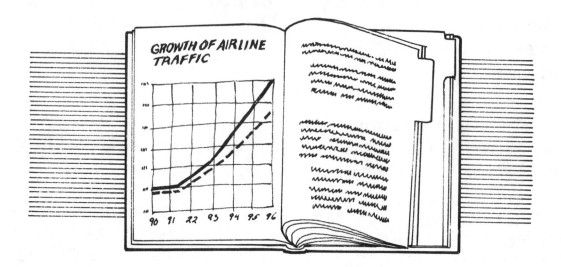

▨ The Log Book

Once the child has set up a tentative plan, the next thing he or she needs to do is set up a **Log Book**. A Log Book is like a diary or a journal. It is a complete record of the child's work.

The Log Book will contain a variety of information such as the planning calendar, lists of possible topics for inventions, information from the library that may be written or photocopied, pictures, information about how to develop the invention, materials needed, and data gathered. Some parts of the Log Book may be printed or written in pencil. Other parts may be written in ink or printed from the computer.

A Log Book should always be present when the child is working on the invention. By the end of the invention, it will probably contain water spots or smudge marks. This is acceptable and represents the child's hard work.

A Log Book should never be rewritten or recopied. Spelling and grammar do not have to be perfect. The child may use any kind of blank book as a Log Book. He or she may wish to include the Log It! pages found in this book. These pages are found throughout the book, and complete details are given on how to use them. If the child is conducting a science fair study, the Log Book will need to conform with the science fair guidelines. In other studies, the child may decide personally what sections to include.

Scientists have been keeping Log Books for a long time. Theirs, like the child's, are not changed or recopied. Scientists have learned much information from reading earlier scientists' work. We build on information from the past.

-Planning Calendar and Log Book continued-

The Log Book serves many purposes. It usually contains the following:

- a daily journal in which to record thoughts, decisions, and reflections
- a time line for the invention study
- background information
- a record of decisions made
- raw data for the test of the invention
- a day-by-day record of what has been done throughout the invention study
- a place in which to keep all the important papers relating to the study

The Log Book should begin the day the study begins. It is important to set up the Log Book in a systematic manner. It should be divided into sections. Use tabs that are purchased or made. The tabs generally used in an invention study are as follows:

1. Daily Journal
2. Planning Calendar
3. Choosing a Topic
4. Background Information
5. Designing the Invention
6. Problem and Hypothesis
7. Setting Up Procedures
8. Test Data
9. Results and Interpretations
10. Conclusion
11. Sharing the Study

Page 10 contains tabs that the child may wish to use for his or her Log Book. If the child wishes to use these tabs, have him or her cut out each tab. The child can then glue each tab to the right side of a separate blank sheet of paper in the Log Book. The tabs should be positioned so that they are easily visible when in the closed Log Book.

Log Book Contents

The Log Book is the story of how the invention was developed. It tells everything that happened from beginning to end. The numbered Log It! pages found throughout this book will assist the child in creating a well-documented Log Book. These pages are to be completed by the child and filed in his or her Log Book. In the bottom right corner of each Log It! page is information indicating in which tabbed section of the Log Book the page is to be filed. Regardless of the age of the child, this book will help him or her create a valid piece of scientific research. Below and on pages 7–8 are the recommended sections the child can use in his or her Log Book when doing an invention. Also included are explanations of each Log It! page that will be a part of these sections.

▓ 1. Daily Journal

This area of the Log Book is where the child records his or her day-to-day thoughts, ideas, reflections, and actions. The child should write about trips to the library or to a museum. Anything that helps in the development of the invention should be recorded in the daily journal.

▓ 2. Planning Calendar

Log It! #1 (page 9) is used by the child to determine the plan and time line for the invention. Categories are listed to assist the child in planning.

▓ 3. Choosing a Topic

A child has many decisions to make when choosing a topic for developing an invention. Chapter 2 deals with how to choose a topic for an invention. Log It! #2 (page 13) is designed to help the child work through this process.

▦ 4. Background Information

AMELIA EARHART

This section helps organize both the search for information and the information collected. Log It! #3 (page 18), *Brainstorming Blast!*, should help the child with his or her search for information about the invention. There are four types of background information. They are historical, importance, factual, and procedural. **Historical information** about early researchers or past research dealing with the invention study is included on Log It! #4 (page 19). Information concerning the **importance** of the invention to the child and to mankind goes on Log It! #5 (page 20). **Facts** and terms about the invention go on Log It! #6 (page 21). The **procedures**, or ways to do the invention study, should be placed on Log It! #7 (page 22). Log It! #8 (page 23) provides the child with a place where he or she can record and analyze each piece of background information that was gathered. Make as many copies of page 23 for the child as needed.

▦ 5. Designing the Invention

This section guides the child through the process of designing the invention study. It takes into account both the possible building modifications of the model and the experimental testing for the model's specifications. Log It! #9 (page 29) is where the child records this design.

▦ 6. Problem and Hypothesis

The problem in an invention is a question about the possibility of creating the invention. It is explained in detail on page 26. The problem should be recorded on Log It! #10 (page 30) along with the hypothesis. The hypothesis is an educated guess as to how the question will be answered and is fully explained on page 27.

▦ 7. Setting Up Procedures

There are two types of procedures to be written. The first type of procedure is for **building the model**. Log It! #11 (page 31) guides the child through this process. It is necessary to include a scaled drawing, step-by-step building procedures (directions), and any needed materials and equipment. The second type of procedure is for **testing the model** to see if it meets the child's specifications. Log It! #12 (page 32) asks the child to write directions for testing the model and to list any materials or equipment needed for the test.

Log Book Contents continued

■ 8. Test Data

The child needs to record the testing data as he or she tries out the invention. This is the experimental portion of the invention study. This testing data should be recorded in the data table on Log It! #13 (page 36). (An example of a data table for Alex's airplane invention is provided on page 35.)

■ 9. Results and Interpretations

The child will need to prepare one graph for each set of testing data that is taken, including successful and unsuccessful models and modifications. Instructions on how to create and prepare a graph are given on pages 37–38. The interpretation of the graphed data and comparisons with specifications should be recorded on Log It! #14 (page 39). (Specifications are explained and mentioned throughout the book.)

■ 10. Conclusion

Log It! #15 (page 42) contains an area in which the child can write a conclusion. This should include reactions to the hypothesis. Does the invention work like the child thought it would in the hypothesis? The child's invention should be compared to similar designs that he or she read about and explained in the background information. The child should include a statement about the importance of the invention to not only himself or herself but also to the world. The child should also include ways in which the invention study could be improved as well as ideas for future studies.

■ 11. Sharing the Study

A visual display should provide an outline of what the child planned to do, what he or she did, and what results were obtained. Spelling and grammar need to be correct, and the display should be pleasing and attractive. Log It! #16 (page 47) aids the child in planning a visual display for a science fair or a classroom presentation.

On page 48, you and the child will find a suggested assessment sheet for an invention. Knowing how a study will be assessed or judged provides a good way for the child to check his or her work. Use it as you complete sections of the invention study and again with the completed study.

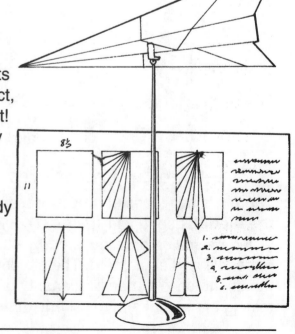

◼ MY PLANNING CALENDAR   Date _____

◼ Essential Information

Rules and Regulations Obtained	YES	NO
Assessment Criteria Obtained	YES	NO
Local Fair or Assignment Due Date		_____
District Fair Due Date		_____
Regional Fair Due Date		_____

◼ Success Calendar

	Planned Date	Date Completed
My Countdown Calendar (1 day)	_____	_____
Setting Up My Log Book (1 day)	_____	_____
Choosing a Topic (2–5 days)	_____	_____
Collecting Background Information (1–3 weeks)	_____	_____
Designing the Invention (1–4 weeks)	_____	_____
Problem and Hypothesis (1–4 days)	_____	_____
Setting Up Procedures (1–5 days)	_____	_____
Getting Materials for Invention (1 week)	_____	_____
Building the Invention (1–2 weeks)	_____	_____
Making a Data Table (1–2 days)	_____	_____
Testing Invention—Collecting Data (1–2 weeks)	_____	_____
Drawing Conclusions (1 week)	_____	_____
Compiling a Bibliography (2–3 days)	_____	_____
Making the Display (1–2 weeks)	_____	_____

Log It! #1
Section: Planning Calendar

Log Book Tabs

Daily Journal

Setting Up Procedures

Planning Calendar

Test Data

Choosing a Topic

Results and Interpretations

Background Information

Conclusion

Designing the Invention

Sharing the Study

Problem and Hypothesis

Choosing a Topic

Chapter 2

In choosing a topic for an invention, the most important thing to remember is that it must be something that is of interest to the child. He or she may want to modify or change a favorite toy. Or, maybe the child is interested in a particular subject such as pulleys or levers and wants to learn more about them. If ideas are needed, have the child think of things that he or she enjoys doing at school, on the playground, at home, or with friends. For example, maybe the child has wondered how airplanes fly as he or she has watched one fly overhead. You might have the child talk with a pilot or an aircraft engineer. You can look for information in the library. Maybe you can pick up some information on the Internet. Below are examples of some possible topics for an invention study. The topic that the child first selects will be too broad for an invention study. It must be narrowed to a particular study question. Possible study questions are also listed below.

Science Category	Possible Topic	Question*
Botany	Plant Growth	Design a new pot for growing plants that allows for good drainage.
Chemistry	Food Spoilage	Make a lunchbox that will keep food fresh for 12 hours.
Earth and Space	Solar Energy	Design and build a solar cooker that will heat water in a closed glass container from 70–80 degrees Celsius in 20 minutes.
Engineering	Building or Modifying Things	Design and build a pinewood derby car or a new toy.
Physics	Electricity Mechanics	Make an electromagnet that will pick up 10 nails. Design an inclined plane that a block will slide down in a certain amount of time.
Zoology	Birds	Build a bird feeder that will attract only cardinals.

It is a good idea to check the rules of your local science fair before beginning an invention study. Review all safety aspects involved in inventing something. Some types of studies may not be allowed while others may require approval before the study can begin. There may be legal, safety, and ethical limitations on studies involving the following: controlled substances, human subjects, pathogenic agents, Recombinant DNA, vertebrate or human tissues, and non-human vertebrates.

It is also important to decide if the invention can be built within the time frame allowed and with available materials and equipment. Help the child narrow the study from a category, to a topic, to a specific question that he or she wants to investigate. The inverted triangle on page 12 shows how to narrow and focus a category to a specific question.

*Each study question should have two parts—a **model** part and an **experiment** part. These parts are further explained at the bottom of page 12.

Narrowing a Topic

The inverted triangle below shows how Alex narrowed and focused his study. Follow the circled words to see the decision-making process.

Category
ENGINEERING

(Child should choose and circle only one topic.)

Topics: (flight,) bridges, automobiles, wind power, building materials, skyscrapers

Possible Questions:

Can a model airplane that stays in the air for a very long time be built?

Can a model airplane that has a 400 cm peak of height in its flight path be built?

Can a paper airplane that will fly a distance of 400 cm be built using an 8½" x 11" sheet of white paper?

(Child should choose and circle only one question.)

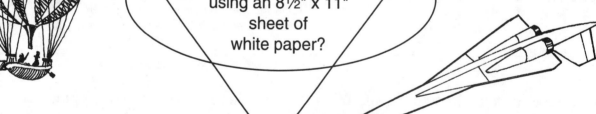

Help the child see that Alex chose "engineering" as the preferred category. Alex listed six topics. Alex chose "flight" as the one he liked the best. Each question that Alex wrote always had two parts—a **model** part and an **experiment** part. Have the child look for each part in the question that Alex selected. The model part is "building an airplane using an 8½" x 11" sheet of paper." The experiment part is the specification of "flying a distance of 400 cm." This is the part that will be tested. Have the child find the model part and the experiment part in each of the other questions Alex wrote above and in the example questions on page 11. Then have the child complete Log It! #2 (page 13) to narrow his or her study.

GETTING DOWN TO BUSINESS

This activity page will help you determine a topic for your invention.

1. Category: _____

2. Topics (List several that you are interested in. Select and circle one.):

3. Possible Questions (List several questions. Be sure each one has a model part and an experiment part. Circle one.):

Log It! #2
Section: Choosing a Topic

Gathering Background Information

Chapter 3

Once the child has chosen a topic and has begun to narrow it, he or she will need to find as much information as possible about that topic. This is called gathering, or collecting, background information. Background information will assist the child in developing the invention design, the problem, the hypothesis, and the procedure. The amount of information gathered will depend on the child and the topic. You will need to help the child gather information that is needed for his or her invention study. Help the child concentrate on pertinent information.

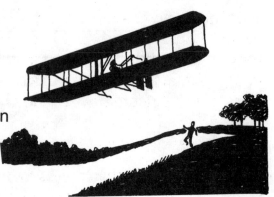

There are four different kinds of background information the child will gather. They are listed and described below and on page 15. Discuss each of these with the child, but do not give the child the Log It! sheets at this time. Log It! #4–#7 (pages 19–22) will assist the child in deciding what background information to gather and where to gather it. These four Log It! pages should be completed after Log It! #3 (page 18) which is described on page 16.

▨ Historical Information (Log It! #4, page 19)

This is what is known about the topic from the past. Did earlier engineers build such models? Is there a famous model that was built? What was discovered? *(Example: Alex would look to see what is known about airplanes and flight by researching such topics as the Wright Brothers and the first flight.)*

▨ Importance Information (Log It! #5, page 20)

This information suggests what is important about the study. Why is studying this topic important to the child and to the world? Could the invention help the child better understand engineering? Could it help others better understand flight? *(Example: Alex would find out how the study is important to the understanding of flight.)*

▨ Factual Information (Log It! #6, page 21)

Factual information consists of what is known about the topic. Facts about the model to be built are needed. This includes facts about the parts that make up the model and what causes the model to go faster, go farther, be lighter, etc. All of this information needs to be gathered. *(Example: In the invention that Alex is working on, Alex would find factual information about the structure of airplanes and what causes one to fly farther than another or better.)*

Gathering Background Information
continued

▨ Procedural Information (Log It! #7, page 22)

Procedural information tells how to build the model and how to test for specifications. First, the child needs information about different designs of the model and the building materials needed. The child also needs to know how to make a scaled drawing and include reference to one. When gathering this information, the child may want to find answers to these questions: What are possible designs? What materials will be needed? Are the materials expensive? *(Example: In Alex's invention, Alex would find books about paper airplanes and how others have made and flown them.)*

The next thing the child needs to do is collect information on how to test to see if the specifications have been met. *(Example: In Alex's invention, Alex would look for information on wind tunnel testing.)*

This may seem like a lot of information to include. However, it is important for the child to have adequate and supportive information for the task. A young child may want to start "doing the invention." This is okay. It gives the child some reality about the task. The child will begin asking questions. Then he or she will begin searching for pertinent facts and other essential background information. The child will probably gather background information throughout the study as questions arise.

The Background Information section will look different from the other sections of the child's Log Book. It can include all or part of the following:

- photocopies of information that the child found,
- tape recordings of people the child interviewed,
- drawings or diagrams made by the child,
- pamphlets received,
- notes the child wrote,
- pages printed from a CD disk on a computer, or
- pages printed from the Internet.

There may be hole-punched pages, pages with glued articles, yellow highlighted sentences, scribbled notes, sketches or drawings, splotches, splashes, and smudges! Each of these types of information is important and adds depth and character to the child's study.

Getting Started

So where and how does the child start gathering background information? A fun and easy way to begin is by brainstorming. This is a great way for the child to find out what he or she already knows about the topic for the invention. Encourage the child to ask friends, parents, teachers, and others to help with the brainstorming. The goal should be to help the child find out as much about the invention topic as possible. Give the child Log It! #3 (page 18) and have him or her write down all the words that he or she and others know about the chosen topic. Be sure the child writes down words that describe or that are related to the topic. *(Examples of words Alex might write include the following: airplane, paper, flight, speed, distance, engineer, lift, aerodynamics, weight, wing size, wing shape, tail, etc.)*

When the child has finished writing down every word that he or she or others can think of, it is time to fit them into one of the four kinds of background information or to cast them aside. When categorizing, the first step is to have the child code the words that he or she wrote on Log It! #3. To do this, have the child scratch out words that do not fit, and then color code the important words using crayons or markers. Words can be color-coded as described below.

- Words relating to history can be circled in blue and recorded on Log It! #4 (page 19).
- Words relating to importance can be circled in black and recorded on Log It! #5 (page 20).
- Words that are facts can be circled in green and recorded on Log It! #6 (page 21).
- Words relating to procedures can be circled in red and recorded on Log It! #7 (page 22).

The child should consider gathering information from many of those listed below.

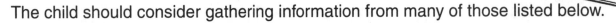

People	Places	Things
family/friends	home	magazines, books, telephone white pages
professionals	government	telephone blue pages
	businesses	telephone yellow pages
	associations	telephone yellow pages
teachers	school	textbooks
librarians	library	books, tapes, Internet
computer experts	computer lab	computer CDs
sales people	bookstores	books, computer CDs
scientists and engineers	museums	apparatus, books, pamphlets
	research labs	books, apparatus
	universities	interviews, laboratories

Recording Background Information

The child has now developed good plans for collecting background information with the help of Log It! pages #4–#7. Reading and recording the most useful information is the next logical step. A *Notes Page* (Log It! #8, page 23) is provided for the child to use to record important information. (Make as many copies of this page as needed for the child. If a photocopier is not available, help the child record only the most useful information.)

Good information is sometimes difficult to identify. A good indicator of useful information is if two different sources repeat the same facts or suggest similar designs. Suggest that the child photocopy useful information. He or she may wish to highlight useful information and glue it to a copy of Log It! #8 (page 23).

▓ Adding Reference Data

In addition to making notes of the materials read, have the child include the reference data. Each Log It! #4–#8 page has a reference, or bibliography, section the child can complete to show where the information was found. The chart below shows examples of needed reference data and how to set it up.

TYPE	AUTHOR/PERSON	SOURCE
Interview	*Name*	*Position, address, phone*
Example	Dr. James Brown	Engineer, Grand Engineering Co. 606 Main St., Peoria, IL 555-1111
Book	*Author*	*Copyright date, Book Title*
Example	J. Jones	1995, <u>Building Paper Airplanes</u>
Magazine	*Author, Article title*	*Magazine date, Magazine*
Example	S. Smith, "How to Build Paper Airplanes"	November 1996, <u>Airplane World</u>
Encyclopedia	*Term or topic*	*Encyclopedia Name, Copyright*
Example	Airplanes	<u>Groliers</u>, 1996

Have the child make a comprehensive list of reference data using all of the references listed on the Log It! #4–#8 pages. The data should be categorized by type and should be in alphabetical order by last name of author. This reference sheet should be placed at the back of the Background Information section in the child's Log Book.

⬛ BRAINSTORMING BLAST!   Date _____

Write down words that describe or that are related to the invention you will be building and testing. Ask parents, friends, teachers, and others to add words to the brainstorming circle. Don't worry about spelling. Write quickly and scatter the words about the circle.

The topic that I am studying is _____.

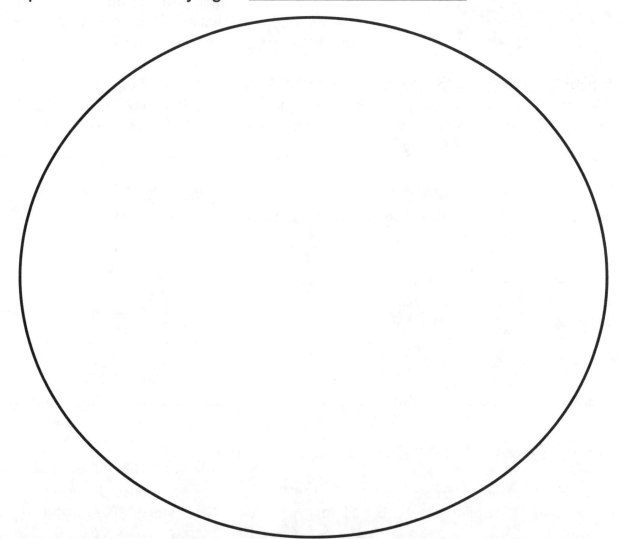

- Circle all the words relating to Historical Information in blue.
- Circle all the words relating to Importance Information in black.
- Circle all the words relating to Factual Information in green.
- Circle all the words relating to Procedural Information in red.

Log It! #3
Section:
Background
Information

Any words that do not seem to relate can be crossed out.

Earlier Scientists and Other History

This page is for information about earlier scientists and/or early scientific information. These early scientists studied inventions similar to the one that you are investigating. Record below the blue circled words from page 18. Define these terms if necessary. Answer the following questions: Did a famous scientist work with this model? What is known about those models?

Past Scientists and/or Past Information:

_____

_____

_____

_____

_____

_____

_____

_____

_____

_____

Possible Sources of Information:

Type (Book, magazine, encyclopedia, Internet, interview, etc.)	Author/Person	Source (Include page numbers or phone number.)

Log It! #4
Section: Background Information

BACKGROUND INFORMATION Date _____

Importance to Me and to Mankind

This page is for recording the importance of the invention to you, others, and other things. Record the black circled words from page 18. Define any necessary terms. Answer these questions: How is the study important to you? Is the study important to other people? Is it important to science? Could this invention lead to a better product? Could this invention make work easier? Could it bring pleasure to others?

Terms   **Importance**

_____

_____

_____

_____

_____

_____

_____

_____

_____

_____

Possible Sources of Information:

Type (Book, magazine, encyclopedia, Internet, interview, etc.)	**Author/Person**	**Source** (Include page numbers or phone number.)

Log It! #5
Section: Background Information

BACKGROUND INFORMATION Date _____

Facts About My Topic

This page is for recording facts about the model and how it works. Record the green circled words from page 18. Define any necessary terms. Answer these questions: What are the most important parts of the model? What causes the model to do what you want it to do?

Terms   **Description**

_____

_____

_____

_____

_____

_____

_____

_____

_____

_____

_____

_____

_____

Possible Sources of Information:

Type (Book, magazine, encyclopedia, Internet, interview, etc.)	Author/Person	Source (Include page numbers or phone number.)

Log It! #6
Section: Background Information

BACKGROUND INFORMATION Date _____

Procedures

Under "Building the Model" below, record the red circled terms from page 18 which relate to the model. Terms such as "scaled model" and "blueprint" might be included. Under "Supplies and Equipment Needed," list any items that might be needed to build the model.

Building the Model

_____

_____

_____

_____

Supplies and Equipment Needed

_____

_____

_____

_____

Under "Testing Specifications" below, record the red circled terms from page 18 which relate to specification testing.

Testing Specifications

_____

_____

_____

_____

Supplies and Equipment Needed

_____

_____

_____

_____

Possible Sources of Information:

Type (Book, magazine, encyclopedia, Internet, interview, etc.)	Author/Person	Source (Include page numbers or phone number.)

Log It! #7
Section: Background Information

NOTES PAGE

Date _____

Reference Data		
Type (Book, magazine, encyclopedia, Internet, interview, etc.)	**Author/Person**	**Source** (Include page numbers or phone number.)

Type of Background Information
Circle One

History   Importance   Facts   Procedures

Recorded Notes

Log It! #8
Section: Background Information

Designing the Invention

Chapter 4

Model D

Once the child has gathered enough background information, he or she is ready to form the study's problem, hypothesis, and design. This is the point at which the child describes exactly what he or she wants to invent and what he or she wants the invention to do. The child uses the background information as a guide to help him or her create the design.

Help the child think of creating the design of an invention as writing a story. A story has these parts, or elements:

      main characters — what the invention story is about

      setting — where, when, and how the story takes place

      plot — what happens when the invention is tested

Scientists do not use these literary terms, but they do use these ideas when designing a study. Scientists use **independent variable** for main character. The setting is referred to as **constant variable** (condition). The plot is called the **dependent variable**. The child's task at this point is to study the background information and select the main characters, setting, and plot for his or her invention—thus, the **design** of the invention.

On page 25 is an example of the design process in relation to Alex's invention. Go over this page with the child before he or she completes Log It! #9 (page 29). Remember, to complete this page, the child must look at the facts and procedures portion of the background information. He or she must organize this information into the proper story elements.

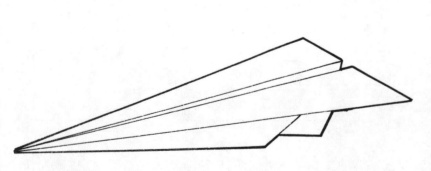

Model C

Designing the Inv

Below is an example of how Alex chose the r
invention study. Alex decided to build a pape
hallway. Alex will use the background informa
the design of the invention.

Alex's Desi

Setting (Cor

Spec

At this point, the child sho
reorganized the backgr
on Log It! #9 (page 2
information on pa
hypothesis, an
components

The child
for the
built

WHERE—The invention will be conducted in a hallway that is 422 centimeters long and 86 centimeters wide. There are no air drafts in the hallway.	**WHEN**—A v invention wi then tested

use weights. If the first plane
does not fly 400 cm down the
hallway after fives tries, Alex
will keep changing the paper
airplane's shape until one is
designed that will fly 400 cm
down the hallway. The design
will be drawn in the form of a
"blueprint" before it is built.

*Main Characters (Independent Variables)	Plot (Dependent Variables)

AIRPLANE SHAPE

A—wide wings

B—

C—

D—

TESTED SPECIFICATIONS

Measure the distance in centimeters that the airplane flew down the hallway. The airplane must fly near the end of the hallway (400 cm).

*Each new shape or modification is its own main character. Therefore, if the first airplane's design focuses on wide wings, wide wings is Main Character A. If this design doesn't work, perhaps the next design will focus on a long tail. This would then be Main Character B.

At this time, help the child complete his or her design on Log It! #9 (page 29).

...fying the Problem

...uld have condensed and
...und information into a design
...9). The child can now use the
...ge 29 to write the **problem**,
...d **procedures** of the study—the three
...hat are part of every invention study.

...begins by writing the **problem**. The problem
...invention is a question that asks what is to be
...and how it will be tested. The problem should be in
...question form and should include the following:

1. what is to be altered (main character, or independent variable—IV)
2. what is being built
3. essential condition (setting, or constant variable—CV)
4. specifications to be tested (plot, or dependent variable—DV)

In Alex's invention, the problem would be written as follows:

Can (1) the shape of (2) an airplane (3) made of paper be altered so that it (4) will fly the 422-cm length of my hallway at least once in five tries?

Alex's **independent variables** are each of the different-shaped paper airplanes needed to be designed in order to accomplish the specified flight distance. Alex made and tested one plane shape at a time. He ended up needing to make four different planes in order to meet the goal of reaching the end of the 422-cm hallway.

The **control** for the model building is the "blueprint" plans used to build the model. The constructed model is always compared to its blueprint intent. In Alex's invention, all testing results are compared to the desired 400-cm distance.

The **dependent variable** in Alex's invention is the distance traveled in centimeters. Alex had to keep creating new airplane designs until he got one that would fly the specified 400-cm length of the hallway.

The **constant variables** are the construction materials and the testing setting. These include the plain white copy paper, the hallway, the meter stick for measuring, five trials of each plane design, and the way the paper airplane is flown during each trial.

Have the child form his or her problem on the top of Log It! #10 (page 30). Encourage the child to make sure that the proposed problem contains each of the four parts listed above.

Writing the Hypothesis

The child is now ready to write the **hypothesis**. The hypothesis is an educated guess as to what the child thinks will happen when the invention is tested. The child is guessing as to what the test results will be. This should not be a wild guess. The hypothesis should be based on the background information and the study design.

Alex decided that this hypothesis should be as follows:

> *I guess that a change in a paper airplane's shape will alter how far it flies down the hallway because air resistance and lift affect flight length.*

The format is always an **I guess . . . because** statement. The **I guess** portion lists the independent variables and dependent variables. The **because** portion of the hypothesis gives a reason for the guess.

The child should now write his or her hypothesis on the bottom of Log It! #10 (page 30).

Setting Up Procedures

▓ Drawing and Writing Directions

Once the child has written the problem and the hypothesis, he or she is now ready to carry out the study design. As was stated earlier, an invention combines the making of a model with an experiment to test that model. First, the child will need to draw a blueprint for the model to be built. Remember that there are two kinds of inventions. One is an **original invention**. For example, Alex is building a never-before-designed paper airplane. This is a new toy and, therefore, involves an original design. The other invention is a **modification** of an already existing thing such as a pinewood derby car. Either type of invention should be drawn to scale so that the child will later be able to build it according to plan.

At this time, the child must write directions for building the invention. These directions must be simple. They can be presented using drawings, labels, and a scale. A **scale** tells what the drawing length represents in terms of the actual completed model. (For example, one centimeter on the drawing might represent five centimeters on the completed model.) You might compare this process to map scales. The child may have studied maps and be familiar with map scales. The child can make a drawing of the model, record procedures for building the invention, and list any materials or equipment needed on Log It! #11 (page 31). If the first design does not meet the specifications, a new Log It! #11 (page 31) will need to be written. One of these pages needs to be completed for each modification that is made. (Alex's example on page 28 will help the child with this process.)

INVENTION— MODEL BUILDING

Date _____

Design ___A___

Make a drawing of your invention on a separate sheet of paper. (Grid paper works very well.) If you are modifying a real object, include a photograph of the real object that you are modifying. Be certain to label parts if necessary. Also make certain to include the scale. Write directions for building the invention in the box on the left. Keep them simple enough so that anyone wanting to could build the same model. List construction materials and equipment which will be needed in the box on the right.

DIRECTIONS FOR BUILDING	MATERIALS/EQUIPMENT NEEDED
1. Obtain a sheet of plain white 8½" x 11" paper. 2. Draw lines on the sheet of paper to show where to make the folds for the airplane. Make certain that the lines are straight and show the exact design of the paper airplane.	several sheets of plain white 8½" x 11" paper Make certain all sheets of paper are alike.

If you have to redesign your invention, be sure to include a r character in the appropriate space on Log It! #9 (page 29) a another copy of this page.

Log It! #1
Section: Se

◼ Testing Specifications

Finally, the child should write directions for testing the model and include any equipment needed to carry out the test. He or she can use Log It! #12 (page 32) for this. *(In Alex's example below, Alex wrote directions for conducting a test to see if his airplane could fly the 422-cm distance of the hallway.)*

INVENTION— MODEL TESTING

Date _____

In the left box below, write the directions for testing the model to determine if it meets the specifications. List any materials and equipment needed to conduct the test in the box on the right.

DIRECTIONS FOR TESTING MODEL FOR SPECIFICATIONS	MATERIALS/EQUIPMENT NEEDED
Stand in the designated hallway. Hold the plane in either hand, bending the arm at the elbow. Release the plane from the same position each time and throw with the same velocity.	For first test, use paper airplane design A.

Log It! #12
Section: Setting Up Procedures

Setting (Constant Variables)

WHERE	WHEN	HOW

Main Characters*
(Independent Variables)

Plot
(Dependent Variables)

A— _____

B— _____

C— _____

D— _____

TESTED SPECIFICATIONS

*Remember, each new shape or modification is its own main character. Therefore, you may have anywhere from one to four or more main characters.

Log It! #9
Section: Designing the Invention

MY PROBLEM AND HYPOTHESIS

The problem for my invention is as follows:

CHECKLIST FOR PROBLEM
☐ altered (IV)
☐ model being built
☐ condition (CV)
☐ specifications to be tested (DV)

Can _____

_____

_____

so it _____

_____

_____

_____

The hypothesis being tested by my invention is as follows:

CHECKLIST FOR HYPOTHESIS
☐ altered (IV)
☐ specifications to be tested (DV)
☐ reason for guess

I guess _____

_____

_____

_____

_____

because _____

_____

_____

_____

_____

_____

Log It! #10
Section: Problem and Hypothesis

FS-62106 Inventions

INVENTION—
MODEL BUILDING   Design

Date _____

Design _____

Make a drawing of your invention on a separate sheet of paper. (Grid paper works very well.) If you are modifying a real object, include a photograph of the real object that you are modifying. Be certain to label parts if necessary. Also make certain to include the scale. Write directions for building the invention in the box on the left. Keep them simple enough so that anyone wanting to could build the same model. List construction materials and equipment which will be needed in the box on the right.

DIRECTIONS FOR BUILDING	MATERIALS/EQUIPMENT NEEDED

If you have to redesign your invention, be sure to include a new main character in the appropriate space on Log It! #9 (page 29) and to fill out another copy of this page.

Log It! #11
Section: Setting Up Procedures

In the left box below, write the directions for testing the model to determine if it meets the specifications. List any materials and equipment needed to conduct the test in the box on the right.

DIRECTIONS FOR TESTING MODEL FOR SPECIFICATIONS	MATERIALS/EQUIPMENT NEEDED

Log It! #12
Section: Setting Up Procedures

Carrying Out the Invention

Chapter 5

The child has just completed many major tasks. He or she has completed the design, problem, hypothesis, and procedures, or directions, for building and testing the invention. Congratulate yourselves! You are doing great! The child is now ready to carry out the study. Building and testing the model is the fun part of any study. Only a few decisions remain.

▒ Decision 1—Samples

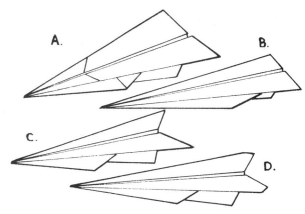

The built model is a sample. How many samples of each model does the child need? In an invention study, this becomes an important decision because the child will have to build variations of the models. If the child is developing a small papier-mâché pot or a recycled paper tray for getting better drainage when growing seeds, then the child will need to make several because other factors can cause the seeds not to germinate. The child can begin building his or her invention at this time. *(In Alex's invention, there was only one sample of each airplane design. This is because each plane can be flown over and over. The only reason that Alex would need to make a duplicate plane is if one of the planes gets destroyed during flight, Alex also needs to build only one plane at a time. If one does not meet the test, then another model can be built.)*

▒ Decision 2—Testing the Invention

On Log It! #12 (page 32), the child should have recorded some decisions about testing the model. Since these plans were made before a model was available, it would be a good idea to have the child look at Log It! #9 (page 29) again to see if it matches Log It! #12. Have the child see if these factors include where, when, and how the test will be done. If it is best to do it outside, the child needs to plan for good weather. Maybe it could be done in the classroom. A scientist might even be willing to share his or her laboratory. *(Alex, as stated before, is doing the experiment in a hallway.)*

▒ Decision 3—Trials

The number of trials refers to how many times the child will repeat the specification test on each model design. The results of the test will be better if the child runs each model test several times. The child has to decide how many trials are reasonable. The child can begin testing his or her invention at this time. *(In Alex's tests of the paper airplanes, each plane would be tested five times. Additional tests would be even better.)*

Making a Data Table

For most invention tests, the child will need to construct a **data table** for the test results. A data table is a place where scientists record the information that they gather from the test. This information is called **data**.

Explain to the child that the data table organizes the information that has been gathered from testing the invention.

Tables are made up of horizontal and vertical lines that are organized into a grid. Explain to the child that horizontal lines go from the left to the right side of the page. The space between them is called a **row**. Vertical lines go from the top of the page to the bottom of the page. The space between vertical lines is called a **column**. Show the child the graphic below.

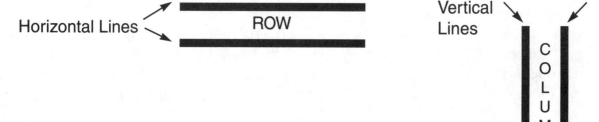

Below are some terms to help the child become familiar with data table components.

- The title of any table is the plot (dependent variable).
- The first row is the main characters (independent variable).
- The next row is the column label. There should be one label for each model that was built and tested (specific independent variables).
- The first column should contain the number of trials, the total numerical value of the trials, and the average of the value of all the trials.
- The remaining columns are where the child records the data.

Have the child look at the data table from Alex's sample invention on page 35. Then the child should record his or her data on Log It! #13 (page 36).

Making a Data Table continued

Sample: Alex's Data Table

Distance the Planes Traveled in Centimeters (Dependent Variable)

Paper Airplane Shapes
(Independent Variable)

		A	B	C	D
T R I A L S	1	101	157	250	410
	2	90	158	300	418
	3	94	151	350	422
	4	86	149	342	422
	5	95	139	325	422
	Total	466	754	1567	2094
	Average	93	151	313	419

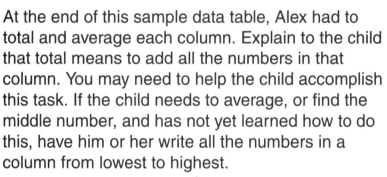

At the end of this sample data table, Alex had to total and average each column. Explain to the child that total means to add all the numbers in that column. You may need to help the child accomplish this task. If the child needs to average, or find the middle number, and has not yet learned how to do this, have him or her write all the numbers in a column from lowest to highest.

1
2
3
4
5
6
7
8
9
10

Then have the child fold the paper at the middle number. This could be used as an average.

Date _____

T					
R					
I					
A					
L					
S					
Total					
Average					

Log It! #13
Section: Test Data

Studying the Results

The child is almost finished with the invention study. He or she has finished testing the invention. All of the data found from the test should be in the data table. Now the child is ready to study the data and see what it all means.

■ Making a Graph

The first part of studying the results is to make a graph. A **graph** is a picture of the results. It is often easier to understand pictures than words or numbers.

The child may wish to make a graph in one of the following three ways:

1. Buy graph paper.
2. Make his or her own graph using a ruler to make the lines. (This is difficult because it is hard to make straight lines that are equal distances apart.)
3. Have someone show him or her how to make a computer graph or chart.

There are some basic rules that need to be followed when making a graph. They are as follows:

- If the child's independent variables are in word form, a bar graph should be made.
- If the child's independent variables are in number form, a line graph should be made.
- The horizontal axis is the line going across, or from left to right, on the page. The horizontal axis is where the independent variable should be placed.
- The vertical axis is the line going from the top to the bottom of the page. The vertical axis is where the dependent variable should be placed.

The graph should have a title. The title should connect the independent and dependent variables. *(In Alex's example, the title would be "The effect of different paper airplane designs on the distance traveled in centimeters.")*

An example of a graph and the information above can be found on page 38.

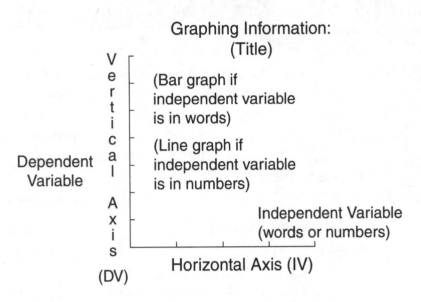

Graphing Information:
(Title)

(Bar graph if independent variable is in words)

(Line graph if independent variable is in numbers)

Dependent Variable

Vertical Axis (DV)

Independent Variable (words or numbers)

Horizontal Axis (IV)

The child should begin numbering at the lower left corner of the graph. If the child determined averages on the data table, this is what he or she should enter on the graph.

To the left is a diagram of how the graph should be set up.

Have the child draw an appropriate graph.

Sample Graph for Alex:

The next step for the child concerning the results section of the Log Book is to interpret, or figure out, what the graph means. How do those bars or lines on the graph compare to the specified value line? Those bars or lines which fall below the specified value line do not meet the

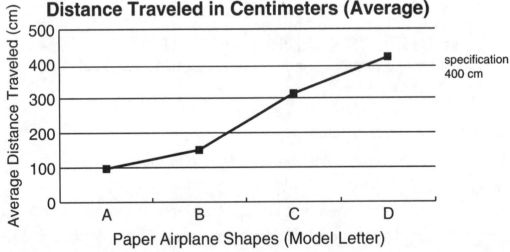

The Effect of Different Airplane Designs on the Distance Traveled in Centimeters (Average)

specification 400 cm

Average Distance Traveled (cm)

Paper Airplane Shapes (Model Letter)

specifications. Those which hit it or are above it do meet the specifications. Below are some questions that may help the child figure it all out.

Which independent variable(s), or main character(s), hits or exceeds what the child specified? *(In Alex's experiment, models A, B, and C were below specifications because they did not travel the distance Alex wanted them to. Model D met and exceeded the specifications as it flew over 400 cm.)*

Give the child Log It! #14 (page 39) to use to write an interpretation of his or her invention results.

◢ INTERPRETATIONS

Date _____

Answer the questions below to help you understand your graph and the results of the invention.

1. Which independent variable(s) or main character(s) reached as high or higher than the specifications?

_____

_____

_____

_____

_____

_____

2. Explain which model(s) met the specifications and why. (For example, how was the successful model(s) different than the others?)

_____

_____

_____

_____

_____

_____

3. Which model(s) was least effective? Why?

_____ _____

_____

_____

_____

_____

_____

Log It! #14
Section: Results and Interpretations

Writing the Conclusion

The **conclusion** is the last part of the invention study that the child must write. It is one of the most important parts. This is where the child explains his or her results. The conclusion involves relating the invention to the entire scientific process that has been developed throughout the course of this book.

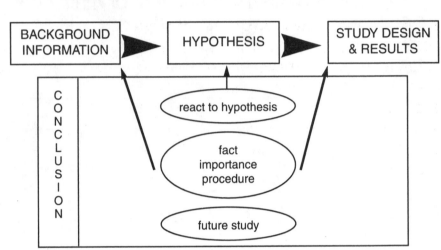

There are five short parts to the conclusion. They include the following:

1. a reaction to the hypothesis and the reason for the reaction,
2. a comparison of the child's study to the background facts and predetermined specifications,
3. the importance of the invention to mankind,
4. checking the procedures—what worked and what did not,
5. a prediction of any similar future studies that are planned.

These parts are further described below and on page 41.

■ 1. Reaction to the Hypothesis

There are three possible reactions. To determine the correct answer, have the child ask himself or herself the following question: "When I compare my study findings to the hypothesis, I see . . ."

- **support**—The model which I chose to use met my specifications. (The child should provide an example or two which supports this choice.)
- **lack of support**—The model which I chose to use did not meet the specifications. (The child should provide an example or two which supports this choice.)
- **uncertain support**—The model which I chose to use met the specifications at some times but not at other times. (The child should provide an example or two which supports this choice.)

(In Alex's invention, the results of model D did match, or support, the hypothesis.)

Writing the Conclusion continued

2. Comparison of the Child's Study to the Background Facts

The child now should revisit the background information. Did his or her invention results agree with what is known about the topic? Was anything new or different from the expected results found during the study? Answers to these questions relate the invention to the factual basis of the original study design.

3. Importance of the Invention to Mankind

Have the child once again look at the background information. He or she should look at the importance of the invention. Are the invention results important to the child? Are they important for one's well-being or for the field of engineering?

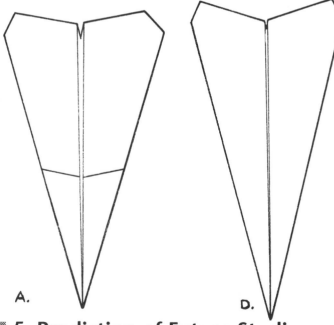

A.

B.

4. Checking the Procedures

The child should look at the invention model designs for strengths and weaknesses. The Log Book may have comments like, "I wish I had done this," or "I wish that I had chosen that . . ." The strengths and weaknesses should be listed with reasons for these comments. You and the child may wish to look at page 48, *Assessing the Study*. This page can serve as a self-evaluation tool for the child as well as an evaluation tool for a mentor or teacher.

5. Prediction of Future Studies

If the child is considering continuing the study, he or she should comment on what would be studied. He or she might like to pursue another aspect of the topic or improve the design.

Give the child Log It! #15 (page 42) so that he or she can write a conclusion containing all five parts.

◢ MY CONCLUSION

Date _____

1. The hypothesis of my study was _____

_____

   Based upon my study, the hypothesis
      is supportive   **is not supportive**   **has uncertain support**
   of my model because _____

_____

_____

_____

2. When I compare my results to facts in the background information,
   I find _____

_____

_____

3. I believe my invention is important because_____

_____

_____

4. In assessing my study design, I have found these strengths: _____

_____

_____

   and these weaknesses: _____

_____

_____

5. I am   **planning**   **not planning**   to continue my invention
   because _____

_____

Log It! #15

Section: Conclusion

42   FS-62106 Inventions

Sharing the Study

Chapter 8

Sharing an invention can be a lot of fun. The child's invention will be interesting for others to see. The child now knows more about this topic than his or her classmates. The Log Book has a lot of valuable information that the child will want to share with others. Informally, the child can share with friends what he or she has done. Formally, the child can share the study in one of three ways. In each of these ways, be certain that the child includes the background information, the problem, predetermined specifications, the hypothesis, the design procedures, the test data, the results, and the conclusion. The three ways to share the study are as follows:

- **Preparing a visual display**—This is a clear and visually-interesting report. Viewers can easily see what the child did and what he or she found out. This is usually required for science fairs.

- **Writing a report**—This is an exacting report written for other scientists. By reading it, others could follow the procedure and repeat the study. The report could also be written for a class.

- **Giving a talk**—This focuses on what is known, what the child did, and what the child found out. A child could give a talk in class.

A title will be needed for the study. Help the child select a short, interesting one. Alex's invention could have titles such as *Planes Soar, The Paper Airplane Challenge, Building the Best Paper Airplane,* etc.

▓ Visual Display

When preparing a visual display, the child needs to realize that science fairs differ in the amount of space that can be used. Most fairs allow a space of at least 60 cm width by 60 cm depth by 120 cm height. This is almost 2 feet by 2 feet by 4 feet. Have the child make his or her display slightly smaller than the display area permitted just to be safe.

Sharing the Study continued

Below are things to consider when making a display:

- Lettering should be easily read from a distance.
- Materials should be arranged to tell a story like a book, left to right, top to bottom.
- The independent variables (main characters) and dependent variables (plot) should be included in each section. The constant variables (setting) are always located in the procedures section and may be included in other sections. The scientific variable terms should be included in the reporting process.
- Color should be used to attract attention. Having each characteristic a different color aids comprehension.
- Photographs or diagrams of the invention should be displayed, not the invention itself.
- Arrangement of information should be interesting, and space should be used carefully.
- The display should be lightweight and be able to be folded flat.
- The display should be able to be quickly and easily assembled.
- The display should have an attached base.

Most displays have three sides, a base, and a title board. Science studies are frequently arranged as seen below. The child can make a 1-, 2-, or 3-sided display. He or she should be certain that it fits within the space allowed. Be sure that the sides are spread open enough so that it can be easily read from the front. Use either a large middle board with smaller side boards or a small middle board with larger side boards.

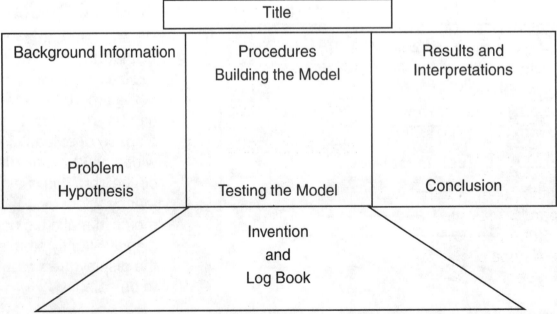

Sharing the Study continued

While the child is planning the visual display, have him or her look at advertisements in newspapers and magazines. This will give him or her ideas on how to use space, color, and letter size. Another place to look is in science centers or zoos. They sometimes have inventions on display. Photographs of science displays are another possible source of information.

Something else the child should check on is whether or not computer printing can be used. It is often much easier for the child to enter information into a computer. Help the child choose a font that can be read easily from a distance. A large font size should be used, and the style should be bold. Have the child use spell check, and help him or her correct any errors and reprint before the information is glued to a colored backing. If a computer cannot be used, the child can use a bold, felt-tip pen for printing. Overhead transparency pens work well. They come in different colors, print boldly, and are easily held. Have the child print on different sizes of unlined sheets, not directly on the posterboard. Lined sheets can be used if needed in the primary grades. Give the child different-sized sheets to write on.

Teacher, office, and art supply stores are great places where the child can find needed materials for the display. Purchase an inexpensive display board. The child will often have choices of color. Have the child use construction paper or thin posterboard for colored backing of the printed material. A good choice for glue is a glue stick.

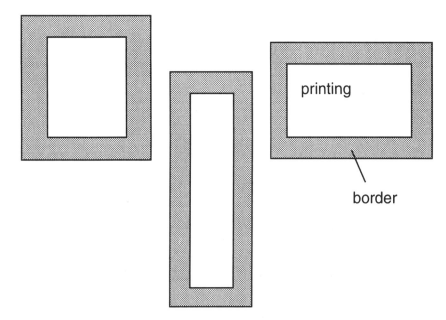

Give the child a copy of Log It! #16 (page 47) to help him or her create a visual display. Be sure that the child has looked at enough advertisements and displays to know how to make good decisions.

■ Report

A report includes all the information the child has on his or her display. However, in a report, this information is shared in greater detail. The invention is shown via photographs in the report. The report is explicit enough so that another person could repeat the same study and check his or her findings with the child's findings. Be sure the child includes the characteristics used and the different appearance of each of them.

■ Talk

A talk includes most of the information that the child has on the display. This is a "show and tell" type of reporting. Tell the child to watch a TV news report of some event to get ideas for his or her talk. A report of this nature usually begins with "how I became interested in the invention." Drawings of the graphs can be used to show the results of the invention. Have the child describe the results. The child can explain what the results mean. The child can end the talk by encouraging others to become interested in the invention and explaining how the invention might be improved in a future study.

■ Good Job!

The invention study has come to an end. Hopefully, the child will have a great feeling of accomplishment. Congratulate him or her for a job well done! The child has worked as a scientist. This may be the first of many inventions that he or she will design or modify. The child has learned how to organize his or her time and how to complete a task. By working with you, the child has developed skills that are valuable when working with other people. These skills will help the child in everything that he or she does.

DISPLAY DESIGN DECISION

Date _____

Below are some decisions you must make when creating a visual display for your invention study.

The title of my display is _____

Sections which will be on my boards include the following:

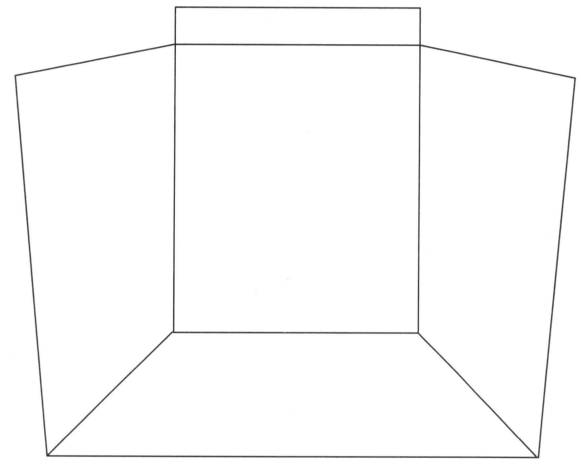

How will the invention be displayed? _____

_____

_____

Color of display board: _____

Color of backing for white, printed sheets: _____

Color of main characters: _____

Log It! #16
Section: Sharing the Study

47   FS-62106 Inventions

Assessing the Study

DISPLAY

Points	Criterion	Explanation	Rating
0 – 2	Easily Viewed	Display faces forward; materials are easily viewed.	_____
0 – 2	Labels	Sections of study design are labeled.	_____
0 – 2	Attractive	Uses color for emphasis; good arrangement; graphics included	_____
0 – 4	Text on Display	Correct spelling and grammar; clear and concise writing	_____
0 – 5	Creative Approach	Evidence of researcher's original input into design	_____

INVENTION STUDY

Points	Criterion	Explanation	Rating
0 – 15	Log Book	A time-task recording of all steps of the study; includes plans, gathered information, design, procedures, data, etc.	_____
0 – 10	Background Information	History, significance, facts, and procedural information are all included.	_____
0 – 5	Problem	Contains question with predetermined specifications	_____
0 – 5	Hypothesis	Includes reason for model design	_____
0 – 20	Procedures	Identifies the variables while describing how the model will be built and tested; indicates how additional tests will be made; uses metric units	_____
0 – 10	Samples and Trials	Appropriate number of model samples and testing trials of controls included	_____
0 – 10	Results	Includes graph showing how well each model achieved the specification; interpretation of graph is given.	_____
0 – 5	Conclusions	Reaction to hypothesis is consistent with test results; includes link to facts, procedures, and significance	_____
0 – 5	Scientific Worth	Thoroughness of plan is apparent; uses dry run; checks for valid study and reliable data; gives possible future study	_____